Rocky Shore

Brian O'Flaherty

REED

CONTENTS

Introduction 4
What makes a rocky shore? 6
The zones of rocky shore life 8
The dry shore and splash zone 10
The high-tide zone 14
In the middle — the mid-tide zone 16
Rock-pool life 18
Low-tide zone — sea edge life 22
The food chain — who eats what? 27
Survival tricks 29
Visiting the rocky shore 31
Glossary 32
Index of plants and animals 33

Introduction

The great New Zealand coast

The New Zealand coastline stretches over 11,000 kilometres, and is one of the longest for a country of its size. The coast varies greatly, and is made up of sandy beaches, bays, estuaries, harbours, sounds and fiords, and open coasts with large areas of rocky shore. On the west coast there are long, sandy beaches interrupted by cliffs and rocky headlands. The east coast generally has numerous bays of sandy or shingle beaches bordered by rocky outcrops with small islets just offshore. About two-thirds of the coastline is hard rocky shore, and the rest of the coastline is soft shores of sand and gravel.

A large diversity of life

Along the coastline, rocky shores can be found near headlands or rocky outcrops, where there are many rocks and boulders. In some areas, a wide, relatively flat shelf may have numerous small pools and crevices.

Rocky shores are found all over the world, but one feature that makes them unique and special as marine habitats is the diversity of life they support. Examples of almost every animal group are found, from seabirds and fur seals to shellfish, anemones and fish. Compared with a

The rocky shore is a place of energy and change.

Iain A. Anderson

Rocky shores are rich in plant and animal life.

sandy beach, rocky shores provide a firm surface for plants to attach to and a variety of sites for animals to live and shelter in: under boulders or in crevices, in rock pools, on and beneath seaweed, out of the water or beneath the waves.

Why are rocky shores important?

- The rocky shore wilderness is unique, diverse and beautiful. It is an area of energy and change, with tides and waves constantly altering the environment. Every day, at low tide, we can visit this rich ecosystem.
- Rocky shores are abundant in life and easily accessible. The shore is most exposed at low tide, so they're great places to explore and study marine life, and important places for scientific research.
- As well as supporting lots of plants and animals, rocky shores are important nursery areas for many fish and crustacean species, as well as being roosting and feeding grounds for birds.
- Sandy beaches or wild rocky shores lie within easy reach of most of our towns and cities, and so the coast is an important part of New Zealanders' lives, for recreation, education, and as a food source (seaweed, shellfish, crabs, sea urchins and fish are taken from rocky shores).

We are fortunate to have easy access to one of the world's richest environments. In just a few metres of rocky shoreline we can find an amazingly diverse array of interesting and unusual creatures.

There's a rocky shore near you!

What makes a rocky shore?

Some shores are cliffs made of hard rocks that are constantly pounded by large waves. Others may be flatter platforms with loose rocks, crevices and hollows, where the waves are gentler. There are also many types in between. How the shore is shaped depends mainly on its exposure to wave action. This varies according to the distance the waves can travel unimpeded and the prevailing wind. Other factors include the rock type and tide effects.

Types of rocks

Rock type varies greatly, from soft sedimentary sandstones and mudstones to harder greywackes and volcanic rocks such as basalt. Each type of rock varies in how it is worn away by wind and water, and this determines whether there will be broken rock, boulders, crevices or hollows and pools. This in turn affects how much shelter the shore provides for animal and plant life.

Waves

The strength and direction of waves also shape the rocky shore. Constant pounding by waves erodes the rocks and creates fissures, boulders, sand and sediment. How strong the waves are influences what species can survive in the area, as well as their abundance.

Tides

The tide's rise and fall is one of the main factors affecting life on rocky shores. During low tide an area of seabed is exposed. How big this area is depends on how steeply the shore slopes into the sea. In different parts of New Zealand, the tide rises from less than 1 metre to 7 metres.

The tide is vitally important for the animals and plants living in the intertidal zone, the area that lies between the low- and high-tide marks. The incoming tide brings nutrients, but at low tide creatures are exposed to sunlight, air and predators. As the tide goes out, pools of sea water are left at various heights above the low-water level.

The constant motion and changing water levels make the shore a difficult place to live, and rocky shore dwellers have adapted to cope with these stressful conditions.

Softer sedimentary rock erodes more easily.

How do tides occur?

Tides are the rise and fall of ocean waters, caused mainly by the gravitational force of the moon pulling the earth's oceans. The sun's gravitational force also affects tides.

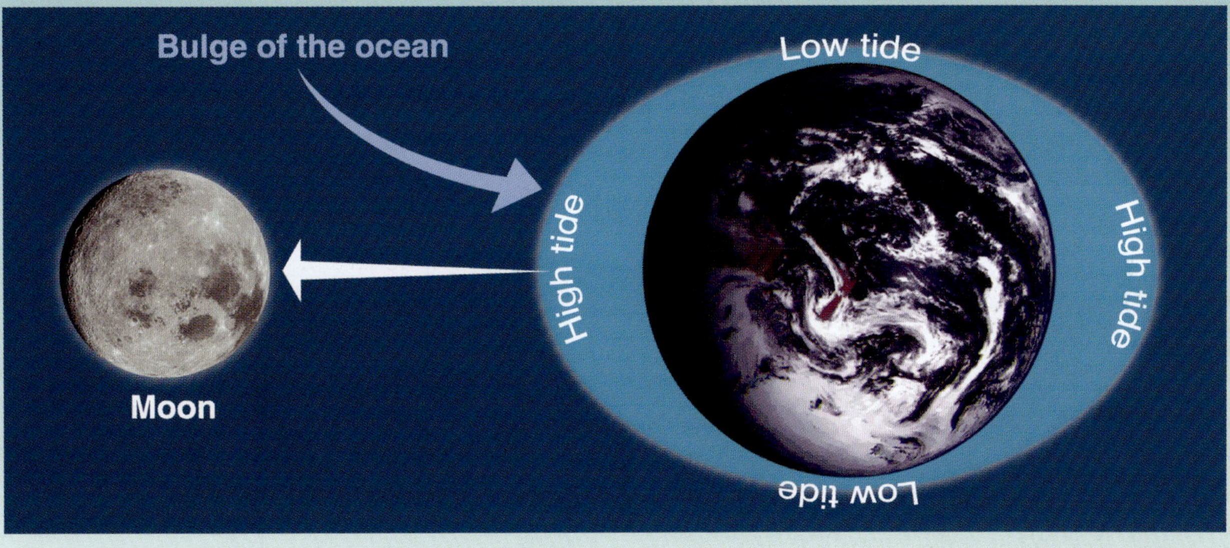

The gravitational force of the moon causes tides.

- Waters on the part of the earth facing the moon are pulled towards the moon, causing a *high tide*. At the same time a high tide occurs on the opposite side of earth. Oceans between these parts of the planet experience a *low tide*.
- When the sun and moon align in space, their combined gravitational forces act on the oceans, creating especially high tides, called *spring tides*. These occur twice a month, shortly after new and full moons.
- When the sun and the moon are at right angles to each other, the moon's gravitational force is reduced by the sun's gravitational force, creating lower than usual tides, called *neap tides*. Neap tides occur twice monthly after the quarter moons.
- High tides occur every 12 hours and 25 minutes, giving a cycle of two high tides and two low tides every 24 hours and 50 minutes. Because the relative positions of the earth, sun and moon are continuously changing, the time and level of the tide varies from day to day.

THE ZONES OF ROCKY SHORE LIFE

A variety of habitats

Weather, water and tides shape the shore, and create a variety of habitats, each with its own survival challenges for plants and animals living there. Common habitats are:

- **Rock platforms**, usually covered and uncovered by the tides. Often the back of the rock forms a cliff, while the ocean edge of the platform steps down into the water. One platform can support many different kinds of plants and animals, because some sections are almost always under water, while other parts are usually dry.
- **Rock pools**, which can vary in size and depth, and provide a constantly wet environment, but experience changes in temperature, salinity (salt content) and oxygen levels.
- **Boulder fields**, where boulders often move, grind together and tumble over, making it difficult for plants and animals to live on or under them.

Zonation

One of the most striking features of a rocky shore is the way that certain plants and animals live at specific tidal levels. The pattern of zones often means the shore is divided into bands of colours, according to the colour of the plants and animals living there.

Rock platform with rock pools.

A boulder shore.

There are various ways of naming these zones, but basically they include the areas from high above the shore (dry all the time) through the splash zone (dry most of the time), to the intertidal zone (wet at high tide, uncovered at low tide) to the sea itself.

The plants and animals in each zone may vary with different climates, the level of the tide, and wave and air exposure, but all show special adaptations to living in this area, and the basic pattern of zonation is recognisable on most rocky shore environments.

So, in the relatively narrow space of the shore, there is an amazing variety of habitats, from dry most of the time to wet all of the time. Where an organism lives on the shore is usually determined by how long it can survive out of the water, its tolerance of temperature changes and defence against predators, and how it competes with other creatures for food and space. Species which normally live in the low-tide zone may live higher on the shore in shade, crevices or pools.

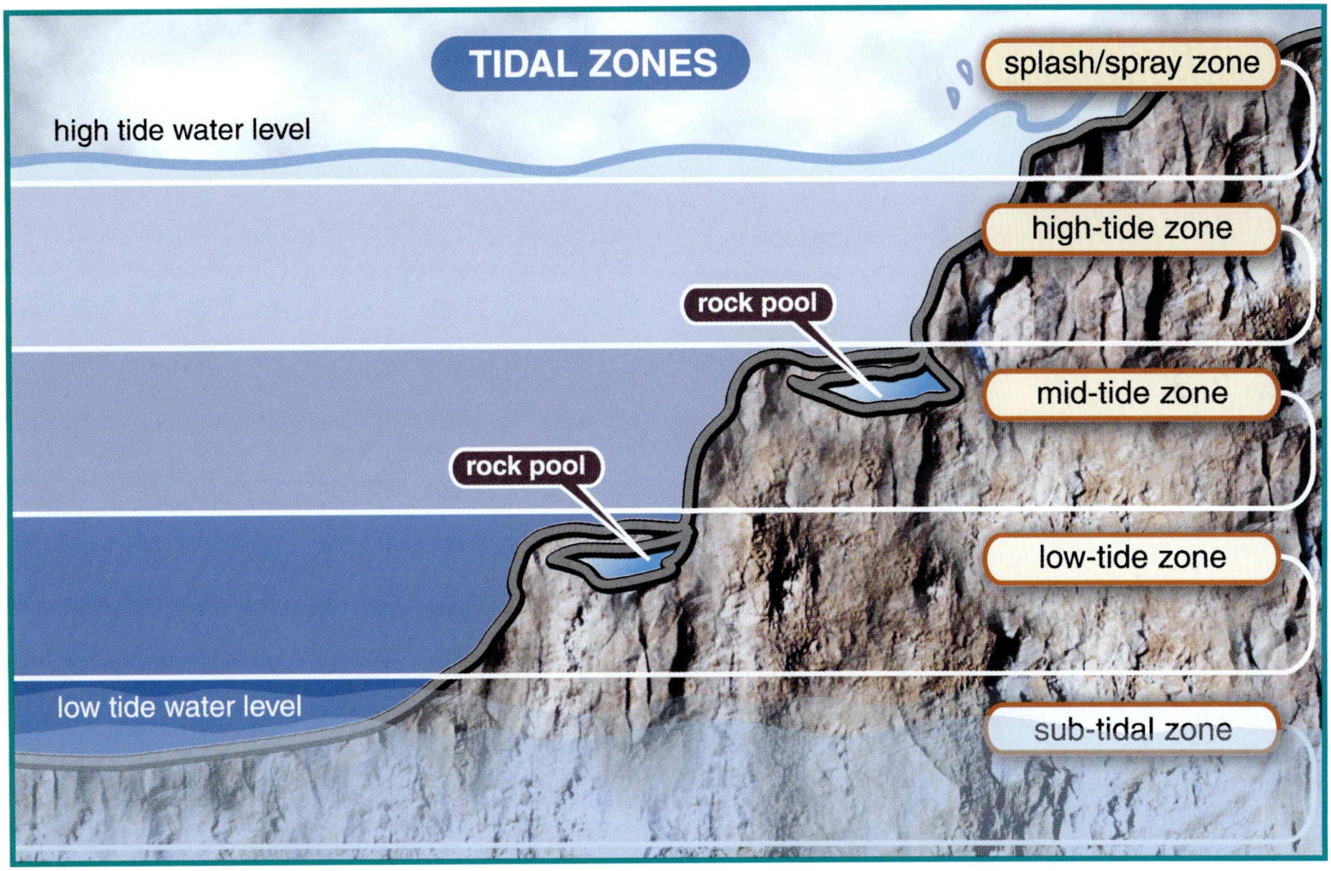

The dry shore and splash zone

Plants

The plants that occur above the rocky shore must withstand salt-laden and drying winds, be able to grab a foothold in often steep, crumbling or rocky ground, and survive in poor soils. Most plants have evolved to survive these harsh conditions — some resemble desert plants. For example, the leaves of **ice plants** have waxy, waterproof coatings to resist absorption of the salt splashed on them, while those of the succulent called **ureure** are tiny and scale-like, helping to slow evaporation. Trees such as **pōhutukawa** and **taupata** have glossy upper surfaces with a waxy coating to resist salt spray. The hardy **leather fern**, often seen growing along the branches of old pōhutukawa trees, has thick, leathery fronds ideal for storing water when it rains. Their upper surfaces are waxy and resistant to spray.

Some vegetation has evolved into different forms to survive the wind's effects. **Ngaio** grows into a dome-shaped tree, whereas **pinātoro** grows in a flat mat. Taupata creeps and twists over rocks with its leaves flattened against the rock surface. The tough spear-shaped leaves of **coastal astelia**, flaxes (like **wharariki** and **harakeke**) and the **rock lily** help withstand wind, salt and water loss.

Beneath the plants and trees of the foreshore and cliffs, rosettes of greeny-grey lichens pattern the dry rocks. These make way for yellow-orange lichen further down the zone. Lichens consist of a fungus and a microscopic algae living together, sharing food and energy to grow. In the area that is wetted by splash and spray, black lichen, which looks like soot, encrusts the rocks.

Like ice plants, many shore plants have waxy, waterproof coatings.

Pōhutukawa — flourishing on the shore

- Pōhutukawa leaves have waterproof, waxy upper surfaces to help withstand the salt spray.
- The leaves often curl under at the edges so spray runs off easily.
- The lower surfaces are covered with fine hairs to prevent moisture loss.
- The tree's tough roots manage to cling to rocks and crumbling cliffsides, and so help to prevent them eroding away.
- The crimson flowers provide nectar for birds.

Birds

Birds that visit the dry shore and splash zone aren't confined to this zone. They can be seen feeding on the rocky shore and in the inshore waters, and often choose the cliffs or vegetation above the shore for nesting, roosting or perching. Birds encountered on rocky coasts include waders such as the variable oystercatcher, gulls, terns, gannets, shags, penguins, herons and the kingfisher.

The **variable oystercatcher**, with its long, bright orange bill, has plumage which varies from black to pied. These birds nest on rocky and sandy coasts. They eat mostly molluscs, worms and crabs, and also small invertebrates and small fish, feeding mainly at estuaries and on sandy shores, and among pools on rocky coasts. Variable oystercatchers are frequently seen in pairs, although they tend to flock in winter months.

David Medway

Variable oystercatcher.

David Medway

Reef heron.

The **white-faced heron** is widespread throughout New Zealand in a variety of habitats, and eats small fish and crustaceans. It nests in tall trees, usually inland.

The **New Zealand kingfisher** migrates to coastal areas during winter months. It can be seen perched on a vantage point, ready to swoop and capture a small crab or fish from a pool.

The wary **reef heron** feeds on small fish, crustaceans and mud crabs, and frequents rocky shores and tidal inlets. It nests on rock ledges, in caves or in the roots of pōhutukawa trees. When fishing, it spreads its wings and has a crouched posture to prevent reflection from the sky. It nests in isolated, inaccessible caves or rock fissures and sometimes beneath exposed roots of pōhutukawa.

Many species of shags live around our coasts. The **spotted shag** dives from the surface for a variety of fish and crustaceans. It is easily seen around our rocky coasts, when, after the breeding season, birds gather together on rocks close to the shore to rest between spells of fishing. They prefer to nest on ledges of cliff faces near deep water, and in alcoves and fissures in volcanic cliffs.

The **large pied shag** can be seen perched on rocks or trees, spreading its wings to dry them. It dives from the surface for live fish, and nests in trees overhanging cliffs, quite often pōhutukawa.

Black-backed gulls live in a variety of habitats, and take a very wide range of food including insects and shellfish, which they hold in their beaks and smash against rocks. At times, they can be seen repeatedly dropping a shellfish on rocks in order to break it open. They nest in a variety of habitats including on coastal rock stacks and cliffs.

Red-billed gulls are slender birds and have grey and white plumage. They are mainly a coastal species, nesting in a variety of locations including on rocky headlands and islets. New Zealand's largest colony of nesting red-billed gulls is on the rocky Kaikoura Peninsula. Their diet is wide and includes planktonic crustaceans, small fish and sometimes the berries of some coastal plants.

David Medway

White-fronted tern.

Some **white-fronted terns** nest on coastal cliff ledges, and rock headlands and islets. This bird feeds mostly on small fishes in inshore waters, caught just below the surface.

The **Australasian gannet** dives to capture fish in the open sea, but sometimes cruises close to breakers, feeding in the shallow water. It nests in large colonies on rock stacks, and can be seen at Cape Kidnappers, Muriwai and Farewell Spit.

The **blue penguin**, the smallest penguin in the world, comes ashore at night to roost or when nesting, and is sometimes seen swimming close to the shore.

THE HIGH-TIDE ZONE

The high-tide zone is exposed to the air, wind, rain and heat between the tides, and so is one of the harshest places on the shore to live. Very few creatures are able to survive here. Animals and plants have to retain water and prevent it being lost by the drying wind and sun. One way to stay moist is to protect all your body with a hard shell. With water sealed inside the shell, some creatures can live out of the water for days at a time. Another way is to group together; this helps to retain water and moist air, and lets the drying wind flow over the clump. Plant growth in this zone is restricted mostly to the algal films grazed by the vegetarian snails.

Periwinkles

The periwinkle is the most common of the grazing snails, rasping algae and filmy black lichen with its tongue (called a radula). They are found from the mid-tide zone up into the splash zone. When the tide goes out they take refuge in crevices and clamp down tightly against the rocks to keep moisture inside. When conditions are excessively dry, periwinkles retreat inside their shells and close their trapdoor (operculum), sealing the moisture inside. They anchor themselves to the rock with a thread of mucus around the rim of their shell.

The high shore is home to millions of tiny periwinkles.

The smaller **common periwinkle** has a banded blue and white shell. They live on all New Zealand rocky shores, sometimes in vast numbers. The larger **brown periwinkle** is more common on the west coast and in southern areas.

Limpets

Limpets are snails that have a cone-shaped shell. You will find them clinging to rocks in the high-tide to low-tide zones or in tide pools. They use a large, flat foot to tightly clamp on to the rock. With their radula they graze on film algae, and as this grows best on wave-swept exposed parts of the shore, they have to hang on tightly. At low tide, they pull themselves tightly against the rock, with rock and shell grinding together.

There are many limpet species in New Zealand. The larger limpets are the easiest to spot. The shell of the **ornate limpet** has radiating black bands, with a row of white pimples

Ornate limpet.

Radiate limpet.

on each. The shell of the **radiate limpet** is larger with radiating lines along the ribs. Many smaller limpets live hidden among barnacles or in cracks and crevices. Some attach themselves to the shells of shellfish and other snails. As they age limpets often become covered with small barnacles.

Barnacles

Barnacles are crustaceans that attach themselves to one spot on the rocky shore and never move about, not even to feed. The barnacle's head is cemented to the rock, so it is not likely it will be washed off by waves. The tough shells are made of several chalky plates, the size and shape of which depend on whether they're exposed to big waves. The white shell of some species reflects the sunlight and helps keep the barnacle cool. At high tide the barnacle opens its plates and, using specialised legs, it filters plankton as it floats past. As the tide recedes, they retreat inside their shell and close the plates tightly.

The large **surf barnacle** covers rocks in the high-tide zone, often in large colonies. **Modest barnacles** are tiny, star-shaped grey-white barnacles that attach themselves to any hard surface.

*These small barnacles (*Chamaesipho columna*) are often joined to form large sheets. They become taller and more hexagonal when crowded.*

Black neritas

On the high shore and around the rock pools in the North Island you'll find lots of these bluish-black, dome-shaped shells. **Black neritas** graze algae off the rocks both night and day. They can shut their shelly operculum and live out of water for a long time. The inside of their shell is shiny white and the animal inside is black.

A huddle of black neritas.

In the middle — the mid-tide zone

The mid-tide zone is a more favourable location than the high-tide zone. Plants and animals here are not exposed to air and sun for as long as those living higher on the shore, and they have longer to feed during each tide. However, better conditions mean greater competition for space and food. Animals must still seek shelter from the elements and predators — inside their shells, or under the cover of seaweed or the safety of crevices and stones. Moist places out of the sun are preferred homes for some for animals in the middle zone, while plants like seaweeds seek the sunlight they need to grow.

Snakeskin chiton.

Chitons

Chitons shelter on or under rocks or in crevices in tide pools. They cling tightly to the rock using suction, and are very difficult to remove. Their hard shell consists of eight overlapping plates that are surrounded by a leathery girdle. The plates and girdle give the chiton more flexibility than a limpet or a snail, letting it mould and seal its shape to the rock and move about without exposing its fleshy foot. Chitons are nocturnal, feeding at night on microscopic plants and algae, which they scrape off rocks with their rasping radula. They usually conceal themselves under rocks during the day. Chitons have eyes but also have light and touch sensors in their shell, allowing them to retreat out of the sun when disturbed.

There are around a dozen chiton species common on New Zealand shores. One that is easily seen is the **snakeskin chiton**. It is named for the banded pattern of the scales on its leathery girdle.

Mussels and oysters

The small black mussel stays small and forms dense beds.

A common feature of the middle shore are beds of **little black mussels** or whitish-grey patches of oysters. Endemic **rock oysters** (in northern areas) and introduced **Pacific oysters** look alike. The mussel beds provide cover for small creatures such as snails, worms and crabs. Mussels and oysters are called bivalves because they have two valves or shells. When covered at high tide the shells open so that they can filter plankton (microscopic floating plants and animals) from the water. At low tide the shells are closed together tightly. This keeps water inside and deters predators. By clumping together, mussels also keep moisture around their shells.

Oysters cement one valve to the rock securely but mussels are anchored by fibres known as byssal threads that the mussel secretes. The number of threads needed to secure the mussel varies depending on how turbulent and exposed its location is.

Neptune's necklace

Neptune's necklace.

The beads (or bladders) in the seaweed called **Neptune's necklace** are used to store water. How big the beads are depends on where the plant grows. Those high on more sheltered shores have bigger beads, storing more water, while those lower on the shore or in rock pools have smaller beads.

Rock crabs

Crabs seek protection from predators and the elements by sheltering in seaweed or under rocks. Turn over a rock on the shore and several crabs will usually scuttle for cover!

There are several crab species encountered on the rocky shore. Two common ones often seen higher on the shore are the **purple rock crab** and the **marble crab**.

- The mature purple rock crab is purple and creamy-white. You see them deep down in crevices high on the shore. They move down the shore at night to feed on snails, limpets and chitons, which they tear from rocks and crush with their pincers. The greenish-coloured juveniles prefer to stay higher up, feeding on barnacles and periwinkles.
- Marble crabs shelter under boulders on the higher shore. They are a shiny purple colour on top and white underneath.

Cool crab facts

- The crab's soft body is protected by its shell, called a carapace. However, at low tide the weight of the carapace can slow the crab down. Movement is easier with the help of water at high tide, when the crab uses its large claws to catch and kill prey.
- As crabs grow too big for their carapace they moult, with the carapace splitting and the crab slowly moving out of the old skeleton, having already grown a new soft carapace. Crabs are very vulnerable to predators at this time.
- Crabs have four pairs of jointed legs and two arms that end in large claws. If a crab loses a leg it can grow a new one.
- Crabs' eyes are carried on stalks, which can be long or short.

Purple rock crab. Iain A. Anderson

Rock-pool life

One way to avoid drying out is to live in rock pools left by the departing tide, but even these are not perfect havens: rock pool water can lose salinity when it rains heavily or through evaporation, become warm in hot weather, and the amount of oxygen dissolved in the water can vary.

Rock pools vary greatly in size and location on the shore, but are found throughout the intertidal zone. However, larger pools lower on the shore experience fewer changes in temperature, salinity and oxygen levels, and so a greater number and variety of marine creatures are found there.

Hermit crabs

Unlike other crustaceans, **hermit crabs** have a coiled abdomen that lacks the tough covering of the rest of their body. To protect this soft part they find empty shells to live in. As they out-

Iain A. Anderson

When the hermit crab outgrows its host shell it finds a new empty shell to live in.

grow the shell they move house to a bigger one. Hermit crabs are a common rock pool resident, and can be seen scavenging for plant and animal remains along the bottom of the pool. Some will filter food from the water. They retreat inside their shell when touched, venturing out again when it is safe.

The most common hermit crab seen in rock pools is grey with bluish markings and hairy fringes.

Sea slugs

The delicate sea slug is also known as a nudibranch, which means 'naked gill' and refers to the feathery gills at the back of its body. Because it has no shell to protect it from predators and keep it moist, the slug relies instead on a bad taste to deter fish and other animals from eating it. Nudibranchs' diet includes various species of sponges.

The **gem nudibranch** lives in shallow water and rock pools in northern areas. The brightly coloured **clown nudibranch** (see photo on rear cover) is common in northern areas, often found in groups feeding on sponges.

Iain A. Anderson

Gem nudibranch.

Sea stars (or starfish)

Sea stars can be found in rock pools or at the water's edge in the low-tide zone. They are not related to fish, so 'sea stars' is a better name for them. They are echinoderms, meaning 'spiny-skinned'.

Sea stars move by drawing water into their arms and pumping it out in ways which lengthen or shorten their thousands of tiny feet. The feet act like suction cups, clasping and unclasping on the rock surface.

Some sea stars don't need a mate. Instead they split in half and grow new arms.

There are many species of sea stars found in New Zealand.

Sea anemone — plant or animal?

Anemones and corals belong to the class called Anthozoa, which is Greek for 'flower animals'. They are animals which grow in forms resembling plants, and are related to jellyfish. An anemone has a sac-like body and tentacles around a single opening. Some can slide about very slowly. There are many species of anemones, found from the high-tide zone to reefs offshore, and in rock pools. They prefer the sheltered sides of rocks as they are protected from the hot sun and drying winds. When out of the water, some species pull in their tentacles and cover themselves with pebbles and shells. Anemones trap small fish and shrimp with their tentacles. Stinging cells on the tentacles then paralyse the prey, which is swallowed whole into the mouth — a fleshy opening in the centre of the ring of tentacles.

Sea anemones can reproduce by dividing in half. Because they can do this, they are able to populate a large area very quickly.

Iain A. Anderson

On the east coast the eleven-armed sea star known as the **spiny star** is common. It feeds on sea urchins, snails and oysters and other shellfish.

The **reef star** is found on exposed west coasts especially where mussels, its primary food, abound.

The five-armed **cushion star** is one of the most common sea stars, and can be seen in pools clinging tightly to rocks or under stones. It is usually greyish-green in colour, but it can range from yellow to purple. On the underside of its arms are tube feet tipped with suckers. The cushion star browses on filmy green seaweeds and feeds on matter found on the sandy bottom of the rock pools. Like other sea stars, it pushes out its stomach through its mouth to cover its prey. Digestive juices are excreted to digest the food.

Spiny star.

Rock-pool fishes

Most rock-pool fishes are small and well hidden, but if you look carefully among the seaweed and stones you will find many different fishes. Around 40 species of fish are commonly found in rock pools in New Zealand, and even a small pool may hold several species.

Even though many fish prefer the protection of rock pools, most will also be found in water to around 10 metres deep. Some reef fishes and open water fishes move into the intertidal zone at high tide to feed, but retreat when the tide begins to fall.

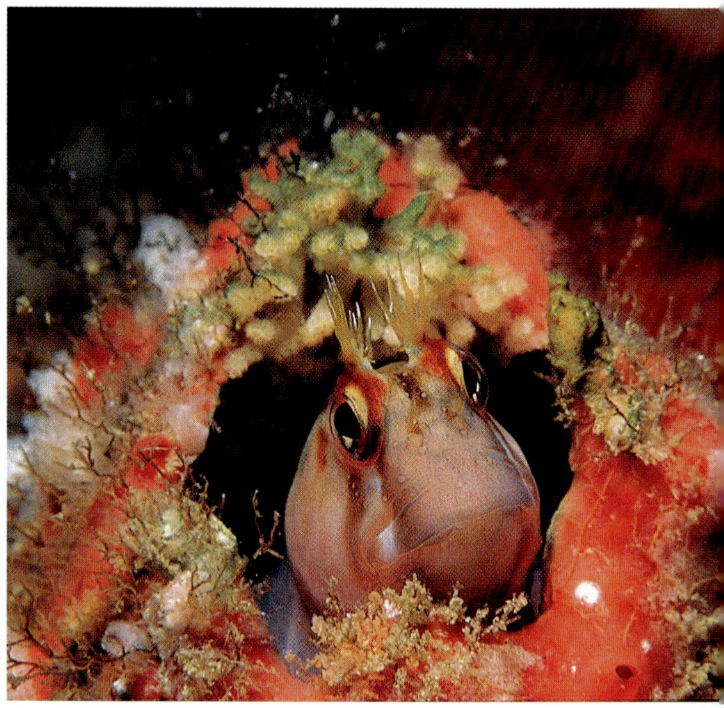

Iain A. Anderson

Crested blenny.

Depending on the species, fish feed on small weed-dwelling and planktonic crustaceans, other small fishes and marine animals. Other

Iain A. Anderson

Blue-eyed triplefin above sponges.

fish are the chief predators of rock-pool fishes, along with birds such as shags, gulls, terns, herons and kingfishers.

Common fishes you may see in the rock pools and intertidal shore include triplefins, rockfishes, blennies, clingfishes, seahorses, pipefishes and weedfishes.

Some of the triplefins ('cockabullies') and blennies are quite distinctive but generally they can be difficult to identify.

Fishes — keeping cool in the pool

Fishes living in rock pools have to withstand changes in temperature, salinity and the amount of oxygen available, as well as coping with drying out and avoiding non-aquatic predators at low tide. They also must tolerate turbulence.

- Being small and often slender or flattened allows fishes to use crevices and other shelter and lessens the effect of turbulence.
- Fishes also shelter in weed and beneath stones, and build their own burrows, or bury themselves in the sand.
- Some species (like rockfishes and clingfishes) protect themselves from drying out by secreting mucus and so can live for periods out of the water between tides, as long as some moisture is present.
- By keeping motionless for long periods and concealing themselves, rock-pool fishes avoid falling prey to birds.
- Their small size and camouflage colouration also helps avoid detection.
- Most have keen eyesight and can move quickly when threatened.
- The shallowness and turbulence of the water in their pools also helps protect them from underwater predators.

Low-tide zone — sea edge life

The low-tide zone, which is uncovered by the tide for only a short time, is the place many rocky shore creatures prefer to live. They are not as exposed to drying air as further up the shore. Some don't have shells and storing water is less important. But, since this is a good place to be, there is more competition for space and more predators, such as crabs and sea stars.

When the tide is at its lowest a great variety of plants and animals is revealed. The rocks are often carpeted with a turf of coralline algae, along with lush red and brown seaweeds, which grow larger in this area than in any of the other zones. Neptune's necklace is also present. This plant life provides cover for animals, which occupy every available space. Seaweed is a vital part of rocky shore life, and an important link in the ocean food chain. Many grazing snails are found here, the cat's eye snail being common. Sea urchins and paua also graze on the seaweed. At the very edge of the sea, kept wet by breaking waves even at low tide, there is usually a curtain of brown seaweed.

Sea urchins

Sea urchins, also known as kina or sea eggs, are echinoderms, like sea stars. Their spines offer good protection from most predators, although they don't always deter fish such as snapper, moki and porae — or humans. The spines are also used for moving about, along with the hundreds of tube feet between the spines. These feet also enable the sea urchin to hold on when grazing, and to catch passing seaweed. The sea urchin's mouth is on the underside, and it has five long, curved teeth. They gnaw away at a lot of the plants they pass over, such as film algae. They also feed on seaweed, particularly kelp and drifting pieces of weed. You can often find sea urchins covered in seaweed that they have saved for later eating. They often let weed and other debris hide them from predators too.

The **common sea urchin** is found all around New Zealand shores.

Iain A. Anderson

Sea urchins can be found in rock pools or below low-tide level.

Pāua

Pāua are not common in pools or under rock ledges at the water's edge any more, because they are such a highly prized seafood, now harvested commercially and also cultivated. They are difficult to see, as seaweed often covers the shell. Pāua grip on to rocks with a strong black foot. They graze on seaweed, generally at night.

Cat's eye snails

The **cat's eye** is named for the sea-green of its thick operculum, which looks like a cat's eye at night. The snail keeps itself moist at low tide by closing this door tightly and sealing water inside the shell. It moves slowly to feed, grazing seaweed off the rocks.

The cat's eye has a beautiful sea-green operculum.

Top shells

There are over 20 species of top shell around New Zealand shores. These grazing snails are found from the middle shore down to the low-tide zone. A common one is the **spotted top shell**, which has a herringbone pattern of black marks over its shell. The **opal top shell** is a conical shape, and lives on brown seaweed in the low-tide zone, as does the **beaded top shell**, which has a brown shell and small pale beads in spirals.

Spotted top shells are a grazing snail of the middle shore.

Sponges

A sponge is a living colony of great numbers of individual animals that build a framework of

Iain A. Anderson

Sponges can be a variety of colours and shapes.

cells. The frameworks have various shapes, from ball-like spheres to branched forms.

Sponges are filter feeders. They draw water through pores in the sponge wall and along a system of corridors. Phytoplankton and bacteria are trapped here, and the filtered water passes out of the larger holes. You can see sponges in deep rock pools, in dark places under ledges or washed up from the sub-tidal level.

Mussels

At low tide at the water's edge, and in the sub-tidal zone, is the large **green-lipped mussel**. This mussel is the one most commonly eaten by New Zealanders, and is more abundant in northern waters. It is cultured commercially, with the mussels attaching to ropes suspended in the water. Smaller **blue mussels** are common further south.

Iain A. Anderson

Green-lipped mussels.

The underwater garden

Seaweed is a primary source of food on the rocky shore, and helps to make it one of the richest areas of life on earth. Seaweed and other algae flourish here because the shallow water lets lots of light through, and there is a steady supply of nutrients brought in by the tide. As well as using the pigment chlorophyll to absorb light during photosynthesis, seaweeds trap different wavelengths of light using additional pigments: red, green or brown. These pigments give the main groups of seaweeds their characteristic colours. Red pigments are better at trapping light, so seaweeds in deeper water are usually red.

The garden of the intertidal zone is filled with an amazing array of seaweeds of various sizes, shapes and colours. Over 1000 species are found around New Zealand shores.

Kelp

Kelp, the largest seaweeds, occupy the sub-tidal zone, and are covered with water nearly all the time. They grow large holdfasts to secure themselves, and their leafy fronds are divided, lowering resistance to the water.

The giant **bladder kelp**, with its distinctive bladder pods that look like large seeds, is one of the fastest-growing plants in the world and forms great masses of weed under water. It is common in colder southern waters.

Bull kelp is another large seaweed, common on west coasts, and in the South Island. It forms great curtain-like buffers at the sea edge, which help to prevent erosion. Air bladders within the blades of the bull kelp act as floats and help keep it near the sunlight at the water surface. Its rubbery holdfast looks like a sink plunger, and its blade may reach 10 metres long.

Ecklonia kelp, common in northern areas, grows from the shore edge to 20 metres or more deep.

Iain A. Anderson

Bull kelp.

Seaweeds — holding, bending, floating

Seaweeds are part of the algae family, and so can live on hard surfaces where plants with roots wouldn't be able to survive.

- Instead of roots, seaweeds have special suckers called holdfasts that adhere to a rock and anchor them in place.
- Seaweeds don't have flowers, or normal stems or leaves. The stem-like part on some seaweeds is called a stipe. It is flexible and can bend when washed by the waves.
- The stipe joins the holdfast to the 'fronds' or blades (the leaf-like parts), which collect the sun's energy and produce the food.
- Seaweeds usually have air pockets or air-filled bladders to keep them near the sunlight at the surface, especially if they live in deeper water. Higher up the shore getting light is not a problem, but drying out is, so the seaweed bladders are used for holding water rather than flotation.
- Not all seaweeds have long blades that float. The blades can be tiny, so the seaweed looks like a soft mat covering the rocks, or like tiny cabbages.

Coralline algae

Reddish-brown clumps of coralline algae 'turf', which is bristly underfoot, often carpet large areas of the lower shore. The stiffness of these plants is due to their skeleton of calcium carbonate, a texture that helps to deter snails and limpets from grazing on them. There are various forms of coralline algae, from pink paint and low mat types to taller, branched specimens where the water is less turbulent. The dense mats provide shelter for numerous small animals, such as snails and crustaceans, and a nursery habitat for the likes of cat's eye snails and whelks.

Cushion star on coralline algae turf.

Seaweed surprises

- A fast bladder! At peak growing times, bladder kelp can grow more than 30 centimetres a day. When fully grown, it can be 25–30 metres long.
- Shampoo with seaweed? Yes, an extract of red **agar seaweed** (from the cell walls) is used as a thickener in many cosmetics and foods. Agar is also used as a culture medium in the laboratory.
- All seaweeds are edible — but some taste better than others. They are full of protein, vitamins and minerals. Try a few beads from the tip of Neptune's necklace — yum!
- Seaweeds have long been part of the diet of various cultures, including Māori.
- We might not enjoy the taste, but like it or not we all need seaweed — it produces oxygen, just like other plants.

The food chain — who eats what?

The real eating action takes place at high tide, as the new tide brings in plankton, and the rising water allows animals to move about more freely, often quite a distance, in search of their prey, without fear of drying out.

Herbivores

Microscopic and large algae along with plankton provide the primary food source on the rocky shore. Snails, chitons, limpets and others rasp seaweed and microscopic plants off the rocks with the radula, their file-like tongue.

Oysters are filter feeders.

Some feeding facts

- Limpets and chiton have iron in the teeth on their radula, which comes from rock ingested as they feed.
- The sea urchin's mouth is underneath. It crawls over seaweed and grinds it up with five specialised teeth.
- Barnacles open their shell plates and extend long, jointed legs that trap scraps of food washing past. Shorter legs then scrape the food off and pass it to the mouth.
- Sea stars can devour bivalves like mussels or cockles. Once they have wrapped themselves around a mussel, they use their tiny tube feet to pull the shells open slightly. Then they push their stomach outside their body and in between the mussel's shells to digest the mussel flesh.
- Like some other carnivorous snails, the **lined whelk** uses its toothed radula to drill through the shells of barnacles, bivalves and even other snails. Once through the shell, it secretes digestive juices and then sucks the creature out of its shell. If you find an empty shell with a hole drilled in it, you'll know who had it for dinner!

Iain A. Anderson

A reef star feeding on green-lipped mussels.

Filter feeders

Animals such as sponges, mussels and barnacles filter plankton and other tiny food particles from the water. Most of these creatures are permanently attached to a rock and so rely on water currents and waves to wash food past them. Filter feeders are spread right across the shore that is covered by water at high tide.

Predators

The grazers and filter feeders become food for others. Carnivores or predators actively hunt their food. Some are very mobile hunters, while others wait for prey to come to them. Predators are found right across the shore, wherever their prey can be found.

Scavengers

After the predators have had a meal, someone needs to pick up the leftovers, and it is the task of scavengers to feed on dead plants and animals. Some crabs scavenge dead and decaying plants and animals. They use their strong pincers to tear their food into small pieces.

Survival tricks

The rocky shore is a difficult place to live, with drying air and sunlight at low tide and battering surf at high tide. There are also great changes in oxygen levels, food availability, salinity and temperature. As well, creatures have to find food and shelter, avoid being eaten, and reproduce. All of these changes and challenges are stressful. Yet some of the largest and most diverse populations of marine plants and animals can be found here. The animals and plants of the shore have developed many survival tricks.

Holding on and hiding out

Plants and animals have developed means to avoid being washed away into deeper water or stranded higher up the beach. Some organisms that are constantly pounded by waves are very tough and can withstand being hit hard. Others are flexible or flat, so they bend instead of breaking when they're hit, or avoid being hit at all. Seaweeds adhere to rocks with their holdfasts.

Many animals avoid sun, drying air and predators such as birds by staying in crevices, under rocks or in their own burrows at low tide. Barnacles and oysters close their valves tightly to avoid drying when the tide goes out. They come out to feed when covered by water.

Some mobile animals have adapted to cope with being washed away. Grazing snails such as the black nerita, cat's eye and top shell all have smooth shells that are quite thick. When washed away they retreat inside the shell and close their operculum, tumbling down the shore. Their rounded shells help them roll along. When they stop, they can re-emerge and hang on to a rock again.

Iain A. Anderson

The seahorse holds on to weed with its tail.

Don't eat me!

On the rocky shore as elsewhere, staying alive means avoiding being eaten as well as coping with the elements. Camouflage works well against predators like birds, fish and crabs, which rely on vision to hunt prey. Colour and shape can be used to blend in with the background. Small size also helps make animals inconspicuous. Rock-pool fishes are an example of this specialisation.

Fending off

If you don't have eyes to see your predators approaching, you have to rely on other methods of defence. Strong shells and their opercula or protective plates do the job for snails, chitons, limpets and bivalves like mussels. The sea urchin's spines, the anemone's 'stingers' and the crab's large pincers all keep potential predators away.

Grow a new part

Some animals, such as sea stars, are able to recover from attacks. Over time, arms lost to predators will grow back.

Cool colour facts

- Most rock-pool fishes rely on colouration to avoid detection by predators (or their own prey).
- **Octopuses** blend in with their background beautifully, and can change the colour and texture of their skin instantly. When threatened, the octopus will squirt out a black ink to momentarily blind any predator. The octopus quickly moves away and disappears!
- **Decorator crabs** attach bits of seaweed or sponge to their backs. If they change location, they can also change their camouflage outfit.
- **Glass shrimp** rely on being transparent to avoid detection. By having no colour, they can remain invisible.

Iain A. Anderson

Octopuses are masters of disguise, changing their colour in an instant.

Iain A. Anderson

The glass shrimp's transparency helps it stay unseen.

VISITING THE ROCKY SHORE

With so much diverse plant and animal life crammed into a small space, the rocky shore offers the chance for exciting discoveries. Planning, preparation and precautions will make your visit safer, more enjoyable and more productive.

Iain A. Anderson

Before you go

- Choose a shore that is accessible and not too wild. Shores composed of sandstone and mudstone are softer and easier on feet and knees than volcanic rock shores.

- Check tide times in your local newspaper. The best time to explore a rocky shore is at low tide, when there is a fairly calm sea, so as to avoid sudden big waves. Aim to arrive at the shore about an hour before low tide for maximum time there. Check the weather forecast too. Strong onshore winds mean bigger waves. Most creatures will try to hide from the sun's heat, so exploring on a cooler or cloudy day may mean you see more.

- Dress for the occasion: sunscreen, sunhat and sunglasses, rubber-soled shoes or aquasocks, clothes that can get wet.

Looking after yourself and the shore

- Watch out for crumbling cliffs if you are beneath them. Stay back from cliff edges.

- Watch your step: barnacles and oysters can slice skin; wet, algae-covered rocks are slippery.

- Keep disturbance to a minimum. Try not to trample delicate plants and animals. Handle animals gently and put them back where you found them.

- Creatures living below boulders are sensitive to disturbance. Try not to move rocks, but if you do, lower them carefully to the same spot to avoid moving or crushing whatever is living on their underside.

Glossary

Algae simple plants lacking true stems, roots or leaves. Includes seaweeds.

Bivalve a mollusc with two shells, connected by a hinge, and closed by muscles.

Carapace the hard shell covering parts of the body of animals such as crabs.

Carnivore an animal that eats other animals.

Chlorophyll a green pigment in plants, used to absorb light during photosynthesis.

Crustacean an animal that has a segmented body with a hard protective shell, such as crabs, barnacles and shrimps.

Echinoderm a spiny marine animal having tube feet and a symmetrical body with radiating parts, such as sea urchins and sea stars.

Ecosystem a community of organisms and their interactions.

Filter feeder an animal that traps food particles, such as plankton, from the water.

Habitat the kind of place in which an animal lives, such as a tide pool.

Herbivore an animal that eats plants.

Invertebrate an animal without a backbone.

Lichen a plant that is a mixture of algae and fungi.

Mollusc an animal with a soft, unsegmented body and often a shell, such as snails and bivalves. An octopus is also a mollusc.

Nocturnal mainly active during the night.

Operculum the 'lid' or 'trapdoor' in various animals, such as snails.

Photosynthesis the process by which plants convert carbon dioxide and water to sugars using energy from the sunlight and producing oxygen.

Plankton microscopic floating plants (phytoplankton) and animals (zooplankton).

Predator an animal that kills other animals for food.

Prey an animal that is hunted by another for food.

Radula a rasp-like tongue with tiny teeth found in many molluscs.

Scavenger an animal that feeds on dead or decaying plant or animal material.

Sedimentary (of rocks) formed by settling mineral and organic fragments.

Turbulence the movement and agitation of the water.

Volcanic (of rocks) formed by the solidifying lava and ash of volcanoes.

Zonation animals occurring in specific places on the shore, creating zones.